Mariana Cassel Meurer
Fernanda Vargas

Yerba mate as a biosorbent for contaminants in water treatment

Mariana Cassel Meurer
Fernanda Vargas

Yerba mate as a biosorbent for contaminants in water treatment

Using yerba mate to prepare activated charcoal

ScienciaScripts

Imprint

Any brand names and product names mentioned in this book are subject to trademark, brand or patent protection and are trademarks or registered trademarks of their respective holders. The use of brand names, product names, common names, trade names, product descriptions etc. even without a particular marking in this work is in no way to be construed to mean that such names may be regarded as unrestricted in respect of trademark and brand protection legislation and could thus be used by anyone.

Cover image: www.ingimage.com

This book is a translation from the original published under ISBN 978-613-9-73082-7.

Publisher:
Sciencia Scripts
is a trademark of
Dodo Books Indian Ocean Ltd. and OmniScriptum S.R.L publishing group

120 High Road, East Finchley, London, N2 9ED, United Kingdom
Str. Armeneasca 28/1, office 1, Chisinau MD-2012, Republic of Moldova, Europe
Printed at: see last page
ISBN: 978-620-7-89615-8

Table of Contents

SUMMARY:

There is currently a great need for measures to improve water quality, which has been damaged by overuse, poor distribution and pollution. Among the technologies available for water recovery, one of the most widely used is adsorption, a process based on the ability of certain solids to remove soluble substances. For this reason, precursor materials with a high carbon content have been tested as adsorbents. The aim of this work is to test the adsorptive capacity of the biomass obtained from the consumed mate herb residue, comparing activation by pyrolysis alone with activation with $ZnCl_2$ and subsequent pyrolysis. The metals tested were Cu^{2+}, Pb^{2+} and Ni^{2+}, which can cause damage to health if ingested in high concentrations. Analysing the results and using the Freundlich isotherms, it was observed that the activated carbon obtained through pyrolysis alone showed greater adsorption capacity for removing the metals studied, when compared to the activated carbon obtained through activation with $ZnCl_2$ followed by pyrolysis, with removal percentages varying between 88.58% and 99.72%, and between 29.11% and 85.06%, respectively. Of the metals studied, copper was the most favourable to remove from water, with almost 100% removal using activated carbon by pyrolysis alone. This process also showed the highest Freundlich constant (Kf) value of 203.0 $mg.g^{-1}$.

Keywords: Water. Adsorption. Metals. Isotherms.

CHAPTER 1

INTRODUCTION

Water is an essential resource for the survival of all living beings, and throughout human history it has become increasingly necessary due to the growing diversity associated with population growth. Biodiversity is greatly affected by any environmental impact on this resource, a common example being the pollution of rivers and lakes by industrial waste and sewage.

Lately, water has been an issue of great concern due to its overuse, poor distribution and organic and inorganic pollution caused by pesticides, waste, mining, industry, infiltration into the soil, structural interventions, inadequate collection and supply, incorrect treatment, among others. As a result, there is a great need to establish appropriate policies and implement effective management systems, which often requires measures to improve the quality of the resource.

Among the technologies used to recover water, one of the most widely used is adsorption, which is based on the ability of certain solids to remove soluble substances. The molecules or ions that are in the aqueous phase, called adsorbates, are adsorbed onto the surface of a solid, called the adsorbent.

The most widely used material for this process is activated carbon, which acts as an adsorbent to remove phenolic compounds and toxic metals from water, but its cost is still very high and it is difficult to regenerate it.

Therefore, combining this concern with the current need to reuse waste, there is a huge search for low-cost, renewable adsorbents that are available in the region in question.

Yerba mate is a tea widely consumed in southern Brazil, Paraguay, Uruguay and Argentina. It is widely used in Rio Grande do Sul to produce the well-known "chimarrão", a hallmark of the state's tradition. The southern region is the largest producer of yerba mate in Brazil, and of the 602,484 tonnes produced in 2014, approximately 600,000 were produced in this region (IBGE, 2014). Given these premises, yerba mate was seen as a promising raw material to be studied as an alternative adsorbent for the treatment of water contaminants.

The aim of this study is to test the adsorptive capacity of activated carbon produced from consumed yerba mate waste, comparing activation by pyrolysis alone with activation with $ZnCl_2$ and subsequent pyrolysis. The metals tested will be copper (Cu^{2+}), lead (Pb^{2+}) and nickel (Ni^{2+}), which can cause damage to health if ingested in high concentrations. To this end, the Freundlich isotherms will be evaluated to check the adsorption capacity of the activated carbons prepared for each of the metals.

This article is divided into 4 sections: theoretical background, which deals with the pollution of rivers by metals, especially in the Rio dos Sinos Basin, as well as the use of isotherms in adsorption tests; methodology, which explains the analysis method used, as well as the parameters used in the analyses; results and analysis, in which the data obtained and the evaluations of the Freundlich isotherms are presented; and final considerations, which contain the conclusions of the work.

CHAPTER 2

THEORETICAL BACKGROUND

2.1 METAL POLLUTION

Some metals, whose ions can be found soluble in water, when in high levels, can contribute to the degradation of water resources (ABREU, 2015). They are generally found in low concentrations, but when ingested in large quantities, they can cause damage to health due to their carcinogenic, teratogenic, mutagenic or even toxic potential. They bioaccumulate in the body, which aggravates the problem of excess metals in the water and increases the damage caused along the food chain (BRAGA et al., 2005).

Normally, due to natural damage, toxic metals are present in low quantities in the aquatic environment, but their quantity can be greatly increased by the discharge of industrial, agricultural or mining effluents. In addition to this problem, other metals, even if present in low quantities, such as sodium, iron, copper and zinc, can be inconvenient to consume because they cause colouring, odour and taste to the water. As a result, there needs to be a concentration limit for the metals cadmium, lead, copper, mercury, nickel, zinc and chromium, which are stipulated by CONAMA Resolution 357 for Special Class water intended for domestic supply with prior or simple disinfection, as well as for preserving the balance of aquatic communities (BRAGA et al, 2005; FEPAM, 2011; SANTOS, 2015).

Due to the degree of pollution caused by metals in the Rio dos Sinos Basin, analysed by the State Foundation for Environmental Protection (FEPAM) from 1990 to 2011, and the resources available in the laboratory to carry out the analyses, the ions Cu^{2+}, Pb^{2+} and Ni^{2+} were chosen to be tested in this work.

The percentage of metals present is very high in the Luiz Rau stream, also popularly called "valão", located in Novo Hamburgo. Lead, chrome, nickel and copper stand out, metals whose main generating sources are metallurgical plants with electroplating, industries found in large numbers in the central region of Novo Hamburgo (FEPAM, 2011).

According to Document DAT-MA No. 0616/2008 (Unidade de Assessoramento Ambiental Geoprocessamento, 2008), stated by Oliveira (2013), the lower section of the Rio dos Sinos Basin, which includes cities such as Novo Hamburgo, São Leopoldo, Sapucaia and Esteio, is the one most at risk due to the large number of industries in the region. In addition, according to Abreu (2015), the Sinos River Basin is generally analysed as "good" or "fair", but in the area encompassing the cities of São Leopoldo and Novo Hamburgo, this concept worsens to "bad", and even to "very bad" in the cities of Portão and Estância Velha, which have around forty tanneries. The author also confirms that, in accordance with CONAMA Resolution 357/2005, lead, copper and nickel, popularly known as heavy metals, have very high concentrations due to the presence of tanneries and electroplating metallurgies in the region. As a result, the water quality exceeds Class 3 and can only be used for navigation and landscape harmony.

2.2 ADSORPTION ISOTHERMS

To measure and evaluate an adsorption process, adsorption isotherms are used, which are generated by the graph that relates the adsorption data and the concentration of adsorbate at equilibrium (YARON; CALVET; PROST, 1996).

For this analysis, the value of q_e must first be determined using Equation 1 (NASCIMENTO, 2014).

$$q_e = \frac{(C_o - C_e)V}{m} \tag{1}$$

Where q_e is the amount of solute that is adsorbed by the mass unit of the adsorbent at equilibrium, in mg.g^{-1} C_o is the initial concentration of the solution, in mg.L$^{-1;}$, and C_e is the equilibrium concentration, in mg.L^{-1} ; V is the volume used in the analysis in L; and m is the mass of the adsorbent in g.

Based on the data obtained from different analyses, a graph of q_e *versus* C_e can be constructed and it can then be assessed whether the adsorption process is favourable or not. A favourable process indicates that the mass of adsorbate retained per mass of adsorbent is high in the case of a low concentration of adsorbate in the liquid phase at equilibrium. When the isotherm is linear, the mass of adsorbate retained by the mass of adsorbent is proportional to the concentration of adsorbate at equilibrium. Finally, when the process is unfavourable, it means that the mass of adsorbate retained per mass of adsorbent is low in the case of a high concentration of adsorbate in the liquid phase at equilibrium. Figure 1 shows the types of adsorption isotherms (NASCIMENTO, 2014).

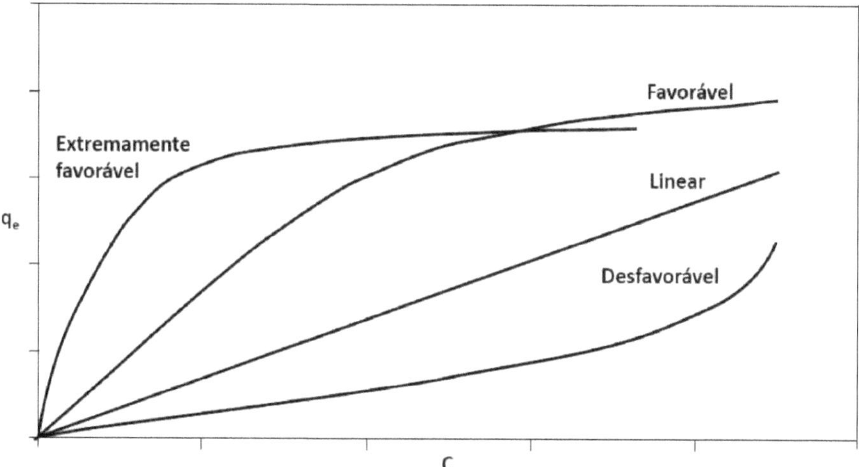

Figure 1 - Possible shapes of adsorption isotherms

Source: Nascimento (2014, p. 26)

In order to assess the adsorptive capacity of certain processes, the Freundlich isotherm uses a logarithmic correlation between the adsorption enthalpy and the concentration of the adsorbate, and is an empirical model that does not predict the saturation of the sites (ABREU, 2013; NASCIMENTO, 2014).

The Freundlich equation in its linear form can be written as shown in Equation 2.

$$\log q_e = \log K_F + \frac{1}{n_F} \log C_e$$

(2)

Where q_e is the amount of solute that is adsorbed by the mass unit of the adsorbent at equilibrium, in $mg.g^{-1}$ nf is the heterogeneity factor, related to the intensity of the process; K_f is the Freundlich constant, in $mg.g^{-1}$, related to the adsorption capacity; and C_e is the equilibrium concentration, in mg.L^{-1} (SANTOS 2015; NIEDESBERG, 2012).

CHAPTER 3

METHODOLOGY

3.1 BIOMASS PREPARATION

To analyse the adsorption of metals using activated carbon obtained from yerba mate waste, three samples of different brands were collected, which were waste from consumed mate. The samples were homogenised and dried in an oven for 24 hours at 110 °C, as adapted from the methodologies of Oliveira et al. (2013) and Cruz (2009). Subsequently, the sample was ground in knife mills, homogenised and sieved to achieve a particle size smaller than *mesh* 170, the smallest particle size possible with the resources used. This value was chosen because several studies have shown that the smaller the particle size used, the greater the adsorptive capacity of the material under study (GONÇALVES et al, 2007; NIEDERSBERG, 2012).

Part of the sample was then chemically activated with ZnCL (zinc chloride) in a 1:1 ratio, according to the methodology of Niedersberg (2012). To do this, 20 g of ZnCL was dissolved in 60 mL of distilled water in a beaker. To this solution, 20 g of the ground yerba mate sample was added. The mixture was left to stand at room temperature for 24 hours. The mixture was filtered through filter paper and the filtrate was dried at 110 °C for one hour, as indicated by Linhares (2013). The dried sample was then pyrolysed at 700 °C in a muffle furnace for 10 minutes. After pyrolysis, the material was washed with a 12.5% HC solution to remove the retained zinc ions and thus unclog the pores. It was then washed with deionised water until it reached a pH of 6.5. Finally, the activated carbon obtained was dried at 110 °C for one hour, cooled and placed in a desiccator.

At the same time, the other part of the yerba mate was not chemically activated, but only physically activated for comparison purposes. The ground yerba mate was dried at 110 °C for one hour and pyrolysed at 700 °C for 10 minutes. It was then washed with deionised water until it reached a pH of 6.5, and then the activated carbon obtained was dried at 110 °C for one hour, cooled and placed in a desiccator. No inert atmosphere was used in the pyrolysis due to limited equipment.

The flowchart in Figure 2 illustrates the process of preparing activated carbon.

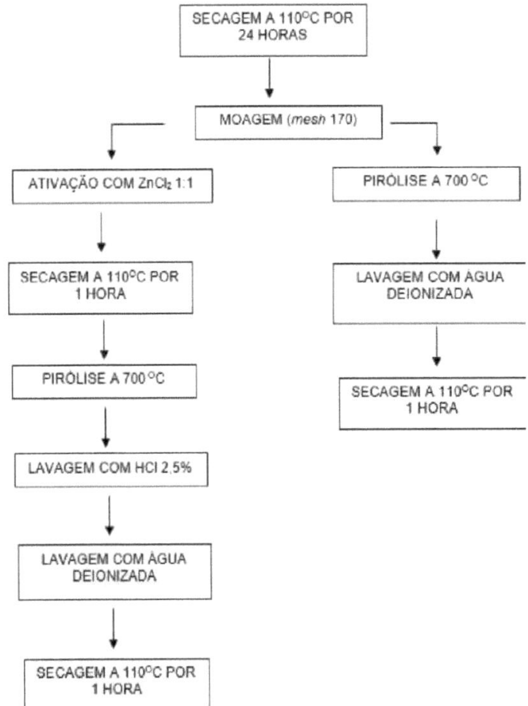

Figure 2 - Biomass preparation process
Source: Author (2016)

3.1.1 Scanning electron microscopy

After preparing the activated carbons as described, they were analysed using scanning electron microscopy (SEM), in order to provide the size of the particles and visualise their surface at a magnification of up to 10,000 times. To do this, the samples were placed on the *stub* and fixed to a carbon strip so that they could be metallised with gold. Once ready, they were photographed at magnifications of 100, 1,000 and 10,000 times on a JEOL scanning electron microscope, model JSM-6510LV, so that a comparison could be made between the charcoal obtained by activation with $ZnCl_2$ and pyrolysis and the charcoal that had only been pyrolysed.

Metal adsorption tests were also carried out, as explained below.

3.2 ADSORPTION ANALYSES

Four solutions of each of the metals to be analysed were prepared, 2, 2.5, 3 and 4 ppm, as suggested by Souza's methodology (2009). For this purpose, solutions of copper II sulphate, lead chloride and nickel chloride were used. The solutions were analysed at this stage by atomic absorption in Varian equipment in order to determine their exact concentrations.

0.08 g of the activated carbon to be tested was added to 100 mL of each solution.

The pH was then adjusted to 5 with NaOH 0.1 $mol.L^{-1}$, adapted from the methodologies of Gonçalves et al. (2007) and Oliveira et al. (2013).

The solutions were stirred at 120 rpm, indicated by Moreira (2010) as a speed that allows effective contact between solid and liquid, for a period of 60 minutes. After this time, the solutions were filtered to remove the adsorbent carbon, and the resulting liquid was analysed by atomic absorption in Varian equipment.

Table 1 summarises the analysis parameters used for each metal.

Table 1 - Adsorption analysis parameters for each metal, using 0.8 g.L^{-1} of adsorbent, particle size *mesh* 170, pH equal to 5, stirring speed 120 rpm, stirring time 60 minutes, and physical activation at 700 °C.

n	Metal concentration (ppm)	Chemical Activation
1	2	$ZnCl_2$
2	2	-
3	2,5	$ZnCl_2$
4	2,5	-
5	3	$ZnCl_2$
6	3	-
7	4	$ZnCl_2$
8	4	-

Source: Author (2016)

The results obtained were used to construct the Freundlich isotherms. In addition, the percentage of metal removal was calculated using the formula shown in Equation 3 (SOUZA et al., 2009).

$$(3) \quad R\% = \frac{(C_i - C_e) \times 100}{C_i}$$

CHAPTER 4

RESULTS AND ANALYSIS

4.2 CHARACTERISATION OF THE MATERIAL BY SCANNING ELECTRON MICROSCOPY (MEV)

The activated charcoal samples were subjected to SEM analyses. The images seen in Figure 3 show the micrographs of the activated yerba mate charcoal subjected to the two types of activation (with and without chemical activation), at magnifications of 100, 1,000 and 10,000 times. In addition, the average particle size was estimated using micrometric scales using the SEM Control Programme *software, in* the images at 1,000x magnification.

Figure 3 - Micrographs of charcoal activated by pyrolysis alone, at magnifications of lOOx (la), l.OOOx (2a) and lO.OOOx (3a) and of charcoal activated with ZnCh and subsequent pyrolysis, at magnifications of lOOx (1b), l.OOOx (2b) and lO.OOOx (3b).

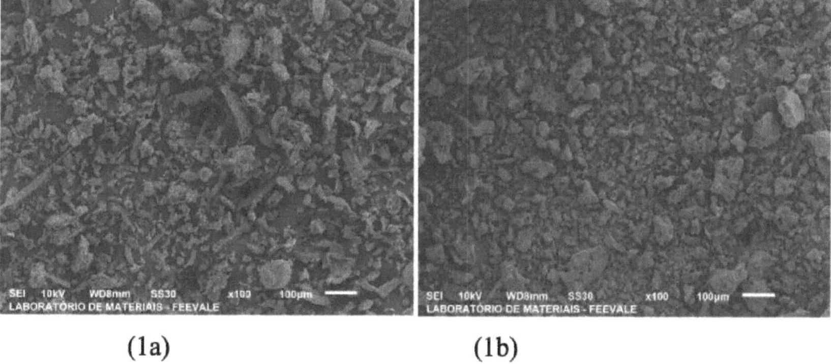

(1a)   (1b)

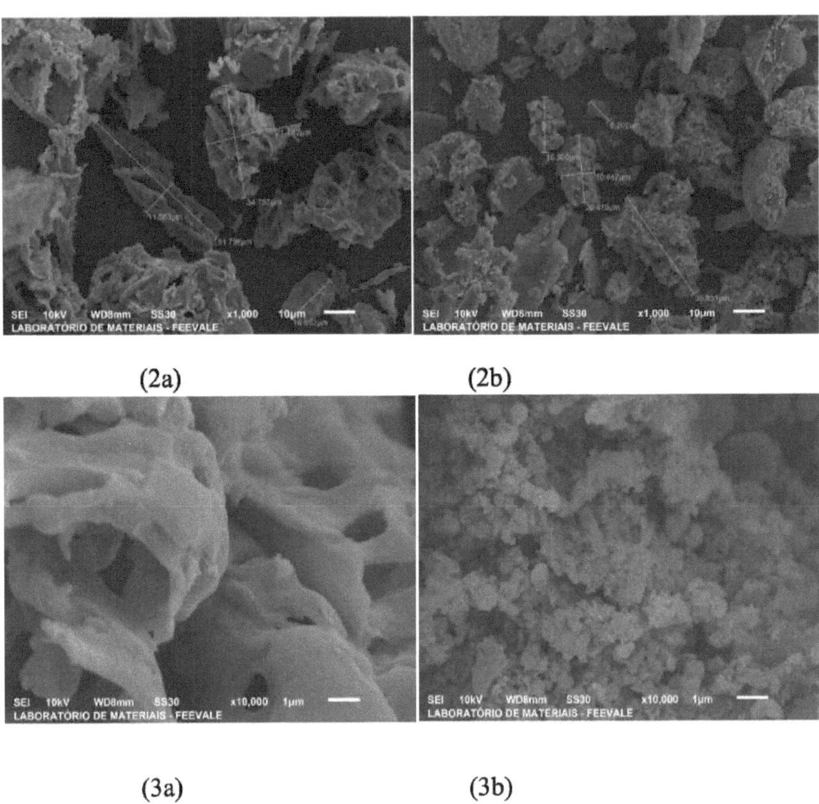

(2a)   (2b)

(3a)   (3b)

Source: Author (2016)

In both cases, a heterogeneous surface was observed, with irregular and disorganised grooves, characteristics observed by Niedesberg (2012), Linhares (2013), and Reis, Lima and Sampaio (2015) who tested chemically activated carbon with $ZnCl_2$, obtained from tung bark, yerba mate and sewage sludge, respectively.

When analysing the images at 100 times magnification, it can be considered that the sample of charcoal activated with $ZnCl_2$ and subsequent pyrolysis has a greater surface area than charcoal activated by pyrolysis alone, since a greater number of particles are observed due to their smaller size. Abreu

(2013) observed greater porosity in sugarcane bagasse treated with $ZnCl_2$ and pyrolysis at 500 °C compared to bagasse treated only with pyrolysis at the same temperature. This difference in size can be seen in the scales of the images obtained at 1,000 magnification. While the particles of the charcoal obtained by pyrolysis alone ranged in size from 16.552 µm to 51.798 µm, the particles of the charcoal activated with $ZnCl_2$ and subsequent pyrolysis ranged in size from 8.202 µm to 35.851 µm.

At 1,000 times magnification, however, it can be seen that pyrolysis-only activated carbon has a surface with many more openings, concavities and exposed pores, which gives it a larger area for adsorbate deposition and thus guarantees a high adsorption rate (NIEDESBERG, 2012).

In the 10,000 magnification images, the greater surface area of the charcoal activated only by pyrolysis compared to the other is even clearer. There is a large surface pore area in the first case, unlike the charcoal activated with $ZnCl_2$ and subsequent pyrolysis, which appears more compact and dense.

4.2 ADSORPTION STUDIES

The metal concentration results, in $mg.L^{-1}$, were obtained in duplicate using an atomic absorption spectrophotometer.

From the data obtained, it was possible to evaluate the Freundlich isotherms for the two different types of activation. To do this, the amount of solute adsorbed per mass unit of adsorbent at equilibrium was calculated in $mg.g^{-1}$ (q_e), using Equation 1. 0.1 L was used for the volume and 0.08 g for the mass. The Freundlich isotherm graphs were constructed using this data, and log C_e and log q_e were calculated, which are the coordinate and abscissa axes on the graph respectively.

The results of the analyses and the evaluation using adsorption and Freundlich isotherms for the copper, nickel and lead metal ions are presented below.

4.2.1 Copper

4.2.1.1 Removal percentage

Table 1 shows the results obtained in the copper analyses of the initial solutions, treated with activated carbon by pyrolysis alone, and treated with activated carbon with $ZnCl_2$.

Table 1 - Concentrations and percentages of copper removal.

Nominal concentration (mg.L$)^{-1}$	Actual concentration (mg.L$)^{-1}$	Average actual concentration (mg.L$)^{-1}$	Concentration after pyrolysed charcoal (mg.L$)^{-1}$	Removal (%)	Concentration after activated carbon with ZnCh (mg.L$)^{-1}$	Removal (%)
1,5	1,57 1,69	1,63	0,0065	99,6	0,96	41,0
2,0	1,78 1,83	1,80	0,0090	99,5	0,87	51,9
2,5	2,29 2,31	2,30	0,0096	99,6	0,85	63,1
3,5	2,85 2,99	2,92	0,0082	99,7	0,84	71,3

Source: Author (2016)

Based on these results, it was observed that the removal of copper by pyrolysis-only activated carbon was very efficient, reaching 99.7% reduction of this metal in the most concentrated sample of 2.92 mg.L^{-1} . The average removal with pyrolysis-only activated carbon was 99.6%, and for activated carbon with $ZnCl_2$ this average was 56.8%, increasing as the initial concentration of copper increased.

Santos (2015) obtained a maximum of 88.2% copper removal using charcoal obtained from pyrolysed orange peels; Ahmad, Kumar and Haseeb (2012) removed 90% of copper using eggshells treated with iron oxide; Silva (2015) showed an average copper removal of 92.8% using chitosan-coated clay minerals. The percentage of copper removal using pyrolysed yerba mate is similar to these values, proving to be even more effective, while yerba mate treated with $ZnCl_2$ showed very low removal capacity.

4.2.1.2   Adsorption isotherms

Based on the results obtained, graphs of the copper adsorption and Freundlich isotherms were constructed for both types of activation, shown in Figure 4. The evaluation parameters were then obtained from the equations formed, which are shown in Table 2.

Figure 4 - Copper adsorption and Freundlich isotherms for pyrolysed activated carbon only (A) and for activated carbon with $ZnCl_2$ and pyrolysis (B)

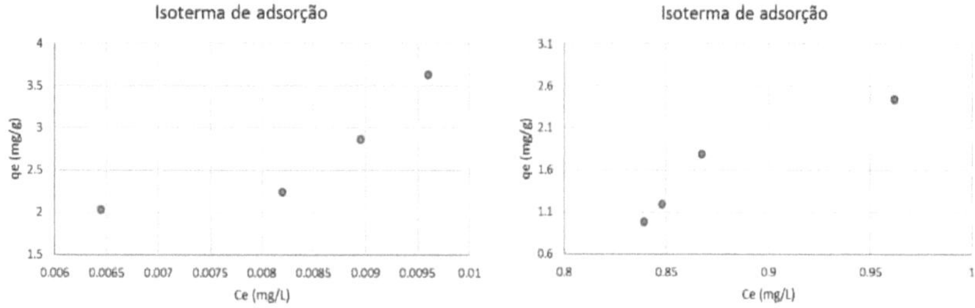

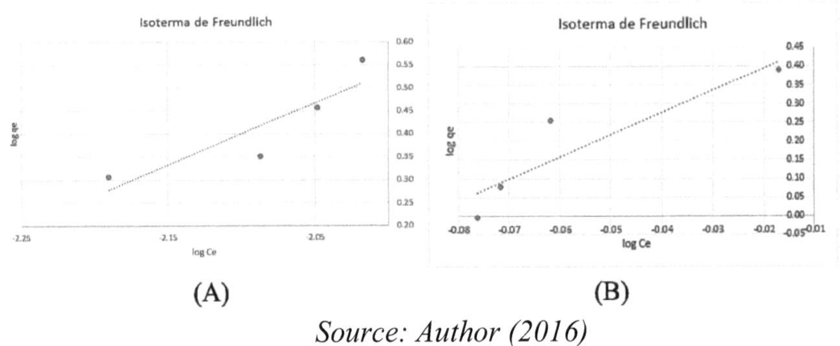

(A)   (B)

Source: Author (2016)

Table 2 - Parameters obtained from the Freundlich isotherms for copper adsorbed by yerba mate activated carbon with pyrolysis only and for activated carbon with $ZnCl_2$ and pyrolysis.

Activation	Freundlich parameters		
	R^2	Df	Kf
Pyrolysis-only activated carbon	0,79	1,09	203,0
Activated carbon with $ZnCl_2$ and pyrolysis	0,83	0,17	3,3

Source: Author (2016)

The R^2 shows that the Freundlich isotherms had a good linearity fit for both charcoal activations, with the highest value of 0.83 being found for charcoal activated with $ZnCl_2$ and pyrolysis.

From the values of nf, which is the heterogeneity factor and relates to the intensity of the process, it can be seen that the adsorption process is favourably physical (nf > 1) using only pyrolysed coal, and that the removal process using activated carbon with $ZnCl_2$ and pyrolysis is favourably chemical (SANTOS, 2015; RUTHVEN, 1964). Ahmad, Kumar and Haseeb (2012) and Abreu (2013), who tested the isotherms for copper removal using eggshell and sugarcane charcoal, found nf values of 1.38 and 2.90

respectively, indicating favourable physical processes.

A much higher Kf value was also observed for the adsorption process using only pyrolysed carbon, confirming, along with the removal percentages, that this is a more suitable and effective process for removing copper from water.

4.2.2 Lead

4.2.2.1 Removal percentage

Table 3 shows the results obtained in the lead analyses of the initial solutions, treated with activated carbon by pyrolysis only, and treated with activated carbon with $ZnCl_2$.

Table 3 - Lead concentrations and removal percentages.

Nominal concentration (mg.L $)^{-1}$	Actual concentration (mg.L $)^{-1}$	Average actual concentration (mg.L $)^{-1}$	Concentration after pyrolysed charcoal (mg.L $)^{-1}$	Removal (%)	Concentration after activated carbon with $ZnCl_2$ (mg.L $)^{-1}$	Removal (%)
1,5	1,78 1,84	1,81	0,16	91,3	0,22	87,8
2,0	1,83 1,84	1,84	0,23	87,3	0,33	82,1
2,5	2,25 2,34	2,30	0,31	86,6	0,37	83,8
3,5	3,75 3,91	3,83	0,34	91,0	0,43	88,8

Source: Author (2016)

As in the case of copper, there was excellent efficiency in the removal of lead by activated carbon with pyrolysis alone, reaching an average of 88.6%. The average removal rate for activated carbon with $ZnCl_2$ was 85.1%, very close to that obtained for pyrolysis-only carbon. No relationship was observed between the removal percentage and the initial lead concentration of the solutions.

The values were close to those of Santos (2015), who obtained removals of between 41.16% and 95% of lead when using charcoal obtained from orange peels, which was the most efficient metal to remove among those studied in his work, which were zinc, cadmium, aluminium, copper, nickel and lead. They were also very close to

by Monteiro, Boniolo and Yamaura (2012), who achieved 90 per cent lead removal when testing adsorption with coconut straw.

4.2.2.2   Adsorption isotherms

Based on the results obtained, the lead adsorption and Freundlich isotherm graphs were constructed, as shown in Figure 5. Next, the evaluation parameters were obtained from the equations formed, which are shown in Table 4.

Figure 5 - Lead adsorption and Freundlich isotherms for pyrolysed activated carbon only (A) and for activated carbon with ZnCh and pyrolysis (B).

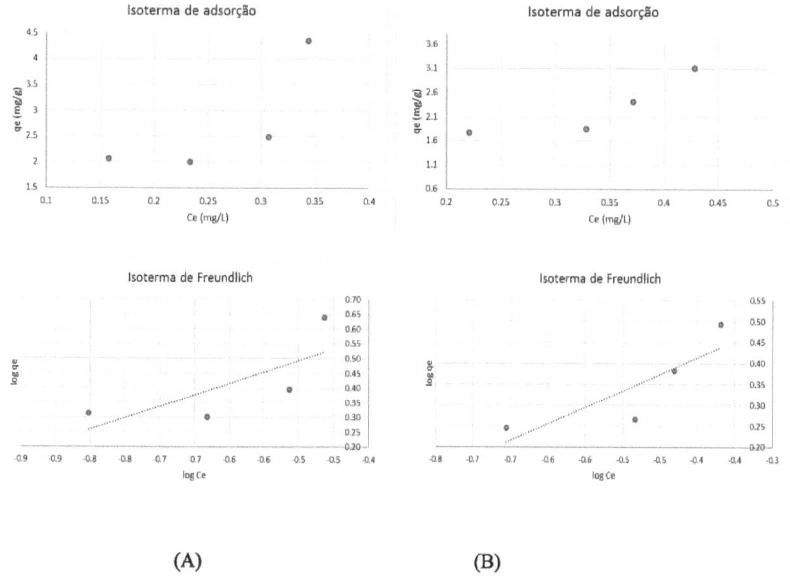

(A)   (B)

Source: Author (2016)

Table 4 - Parameters obtained from the Freundlich isotherms for lead adsorbed by yerba mate activated carbon only pyrolysed and for activated carbon with $ZnCl_2$ and pyrolysis.

	Freundlich parameters		
Activation	**R^2**	**Df**	**Kf**
Pyrolysis-only activated carbon	0,56	1,29	7,60
Activated carbon with ZnCh and pyrolysis	0,73	1,26	5,37

Source: Author (2016)

For lead, the R values[2] show that the Freundlich isotherms had a good linearity fit, although not as good as for copper. The nf values were very close, both greater than 1, indicating a favourable physical process, as obtained by Santos (2015), who obtained nf values above 1 for lead adsorption using pyrolysed orange peel at different temperatures.

From the K values$_f$, we can see a slightly higher value for the isotherm of the pyrolysis-only carbon, of 7.60 mg.g^{-1}, when compared to the activated carbon with ZnCl$_2$ and pyrolysis, with K$_f$ of 5.37 mg.g^{-1}. This difference in adsorption capacity was small, which can be seen in the low difference in the average lead removal in both activations, which were 88.6% and 85.0%, respectively. When testing lead removal with coconut straw, Monteiro, Boniolo and Yamaura (2012) obtained Kf equal to 3.337 mg.g^{-1}, lower than the values obtained with both yerba mate activations.

4.2. 3Nickel

4.2.3.1Percentage of removal

Table 5 shows the results obtained in the nickel analyses of the initial solutions, treated with activated carbon by pyrolysis only, and treated with activated carbon with ZnCl$_2$.

Table 5 - Nickel removal concentrations and percentages.

Nominal concentration (mg.L)$^{-1}$	Actual concentration (mg.L)$^{-1}$	Average actual concentration (mg.L)$^{-1}$	Concentration after pyrolysed charcoal (mg.L)$^{-1}$	Removal (%)	Concentration after activated carbon with ZnCl$_2$ (mg.L)$^{-1}$	Removal (%)
1,5	1,59 1,60	1,59	0,072	95,5	1,05	34,0
2,0	2,22 2,13	2,18	0,13	94,3	1,52	30,1
2,5	2,37 2,36	2,37	0,22	90,7	1,62	31,6
3,5	3,44 3,42	3,43	0,31	90,9	2,72	20,7

Source: Author (2016)

As in the case of copper and lead, the nickel removal results were more efficient in the treatment with pyrolysed coal only, when compared to chemical and physical activation. The average removal rates were 92.8 per cent and 29.1 per cent, the greatest variation found between the two, considering the metals studied. In both types of activation, there was a tendency for the percentage of removal to decrease as the initial nickel concentration increased, a fact also observed by Santos (2015).

The maximum percentage of nickel removal found by Moreira (2010) with ground pecans was 46.2%, which is intermediate to the values found in this study. Santos (2015) obtained a maximum removal of 74.2% with treated orange peel, which is lower than the 92.8% average nickel removal obtained when the adsorption process was carried out with pyrolysed yerba mate charcoal alone.

4.2.3. 2Adsorption isotherms

Based on the results obtained, the graphs of the nickel adsorption and Freundlich isotherms were constructed, as shown in Figure 6. The evaluation parameters were then obtained from the equations formed, which are shown in Table 6.

Figure 6 - Nickel adsorption and Freundlich isotherms for activated carbon with pyrolysis only (A) and activated carbon with ZnCh and pyrolysis (B)

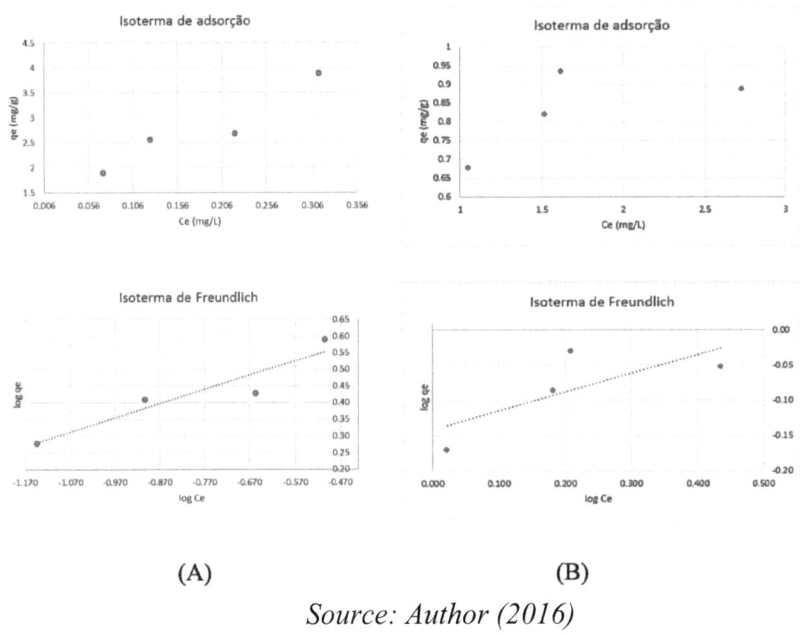

(A)   (B)

Source: Author (2016)

Table 6 - Parameters obtained from the Freundlich isotherms for nickel adsorbed by yerba mate activated carbon only pyrolysed and for activated carbon with ZnCh and pyrolysis.

	Freundlich parameters		
Activation	R^2	Df	Kf
Pyrolysis-only activated carbon	0,88	2,34	5,89
Activated carbon with $ZnCl_2$ and pyrolysis	0,55	3,72	0,72

Source: Author (2016)

As with lead, the R values[2] show that the Freundlich isotherms had a good linearity fit, although not as good as for copper. Checking the Freundlich isotherms, it can be seen that the values of the heterogeneity constants were both greater than 1, once again indicating a favourably physical process. When comparing the values of K_f, a significantly higher value can be seen for the isotherm of the pyrolysis-only charcoal, 5.89 mg.g^{-1} , when compared to the activated charcoal with $ZnCl_2$ and pyrolysis. This figure is very close to the K_f values found for lead adsorption, whose removal percentages were also very close to the removal of nickel with pyrolysis-only charcoal, at approximately 90%.

Santos (2015) found nf values for nickel between 3.33 and 19.28, and K_f values lower than those found in the present work when testing nickel adsorption using pyrolysed orange peel at different temperatures. Moreira (2010), on the other hand, obtained a lower nf value and K_f of 0.70 mg.g^{-1} using pecan biomass, showing very similar efficiency to the removal of nickel using mate coal activated with $ZnCl_2$ and pyrolysis.

4.2.4 Comparative analysis of the adsorption of the metals studied

Based on the results obtained and analysed, a comparison was made to assess which of the metals - copper, nickel and lead - achieved the most effective removal. To make it easier to see, the graph in Figure 7 shows the percentage removal of metals using yerba mate charcoal activated by pyrolysis alone and charcoal activated with $ZnCl_2$ and subsequent pyrolysis.

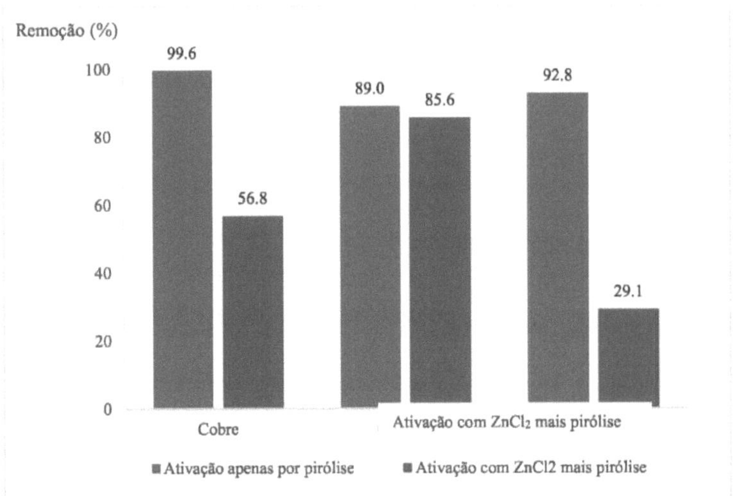

Figure 7 - Graph of the removal percentages of the metals studied
Source: Author (2016)

The decreasing order of removal found in the tests with only pyrolysed coal was: Cu > Ni > Pb. The removal percentages ranged from 89.0% to 99.6%. For the tests with activated carbon with ZnCh and pyrolysis, the increasing order of removal was: Pb > Cu > Ni, with percentages between 29.1% and 85.6%.

According to McBride and Blasiak (1979), one of the factors influencing the preference of one metal over another in the adsorption process is the atomic radius, which is directly related to electronegativity. The smaller the atomic radius, the greater the

electronegativity of the element, i.e. the greater the force of attraction on the electrons in a bond (COX, DOUDNA, O'DONNELL, 2012).

Thus, the order of selectivity of the metals studied should be: Pb > Ni > Cu.

Another factor that affects the preference for adsorption between metals is the free energy of hydration, which the higher it is, the greater the tendency for the metal to remain in solution, and the lower it is, the easier it is to adsorb (OUKI and KAVANNAGH, 1997; SANTOS, 2015; SHERRY, 1969).

Thus, the order of theoretical selectivity would be: Pb > Cu > Ni.

Therefore, the results found in the process of adsorption with activated carbon with $ZnCl_2$ and pyrolysis were within the expected range of the literature cited, observing the order of preference indicated by the free energy of hydration.

On the other hand, the order of selectivity found for the adsorption process using only pyrolysed coal showed a discrepancy in relation to the data in the literature, but the metal removal values were all above 89%, demonstrating the effectiveness of the process.

CHAPTER 5

FINAL CONSIDERATIONS

This article presented data on the removal of metals from water using the adsorption method with activated carbon obtained from yerba mate residue from consumed mate. It was found to be an efficient method for removing copper, lead and nickel from water.

Comparing the activation methodologies (pyrolysis/$ZnCCl_2$ and pyrolysis), it was concluded that the use of charcoal obtained through the pyrolysis process alone has a higher adsorption capacity for the removal of all the metals studied, and could be a potential adsorbent for their removal in industrial processes and treatments. The average removal percentages for copper, lead and nickel in the process that used only pyrolysed coal were 99.7%, 88.6% and 92.8%, respectively. The adsorptive process using yerba mate coal activated with $ZnCl_2$ followed by pyrolysis had removal percentages of 56.8%, 85.1% and 29.1% for copper, lead and nickel, in that order.

It was also concluded that of the metals studied, copper is the most favourable to be removed from water using activated carbon obtained from yerba mate, since almost 100% of the metal was removed from the prepared solutions when treated with activated carbon by pyrolysis alone.

As for the adsorption tests using yerba mate biomass activated with $ZnCl_2$ followed by pyrolysis, the best removal results were seen for lead, followed by copper and nickel.

Finally, the Freundlich isotherms showed that the copper removal process using activated carbon obtained from yerba mate is the most favourable, since it showed the highest K value$_f$ of 203.002 mg.g^{-1} , considering all the tests carried out.

REFERENCES

ABREU, C. A. **Geochemical Distribution of Heavy Metals and Other Chemical Elements in Groundwater of the Sinos Basin, RS.** 2015. 93f. (Postgraduate Programme in Geosciences) - Federal University of Rio Grande do Sul, Porto Alegre, RS, 2015.

ABREU, M. B. **Preparation of Activated Carbon from Sugar Cane Bagasse and Its Application in the Adsorption of Cd(II) and Cu(II).** 2013. 53f. (Degree in Chemical Process Technology) - Federal Technological University of Paraná, Apucarana, 2013.

AHMAD, R.; KUMAR, R.; HASEEB, S. Adsorption of cu^{2+} from aqueous solution onto iron oxide coated eggshell powder: evaluation of equilibrium, isotherms, kinetics, and regeneration capacity. **Arabian Journal of Chemistry.** v. 5, p 353-359,2012.

BRAGA, B. et al. **Introduction to Environmental Engineering.** 2 ed. São

Paulo, SP: Pearson Prentice Hall, 2005. 305p.

COX, M. M.; DOUDNA, J. A.; O'DONNELL, M. **Molecular biology: principles and techniques.** Porto Alegre: Artmed, 2012. 916p.

CRUZ, M. A. R. F. C. **Utilisation of Banana Peel as a Biosorbent.** 2009. 67f. (Master's Degree in Natural Resources Chemistry) - Universidade Estadual de Londrina, Londrina, 2009.

FEPAM-RS. State Foundation for Environmental Protection. **Environmental Quality:** Guaíba Hydrographic Region. Water Quality of the Sinos River Basin. Porto Alegre, RS, 2011. Available at: <http://www.fepam.rs.gov.br/qualidade/qualidade_sinos/sinos.asp>. Accessed on: 20 March 2016.

GONÇALVES, M. et al. **Production of Charcoal from Yerba Mate Waste for the Removal of Organic Contaminants from Aqueous Media.** Lavras, MG: Universidade Federal de Lavras, v. 31. n. 5. p. 1386-1391. set./out. 2007.

IBGE - Brazilian Institute of Geography and Statistics. **Production of Permanent and Temporary Crops.** 2014. Available at: < www.ibge.gov.br>. Accessed on: 28 March 2016.

LINHARES, Bruno Chaves. **Production of Charcoal from Yerba Mate Waste and Its Use as an Adsorbent in the Removal of Organic Pollutants from Aqueous Solutions.** 2013. 39f. (Master's Degree in Process Engineering) - University of Santa Maria, Santa Maria, RS, 2013.

McBRIDE, M.B.; BLASIAK, J.J. Zinc and copper solubility as a function of pH in an acid soil. **Soil Science Society of American Journal,** v.65 n. 3, p.866-870, 1979.

MONTEIRO, R. A.; BONIOLO, M. R.; YAMAURA, M. **Use of coconut fibres in the biosorption of lead in industrial wastewater. Nuclear and Energy Research Institute.** 2012. 15f. V Congress of Initiation in Technological Development and Innovation of UFSCar - Federal University of São Carlos: São Paulo Capital, 2012.

MOREIRA, D. R. **Development of Natural Adsorbents for the Treatment of Electroplating Effluents.** 2010. 79f. (Master's Degree in Materials Engineering and Technology) - Pontifical Catholic University of Rio Grande do Sul, Porto Alegre, 2010.

NASCIMENTO, R. F. et al. **Adsorption: Theoretical Aspects and Environmental Applications.** Fortaleza: University Press, 2014. 256p.

NIEDESBERG, Carolina. **Adsorption Tests with Activated Carbon Produced from Tungue Bark *(Aleurites fordii)*, Waste from the Oil Production Process.** 2012. 65f. (Master's Degree in Environmental Technology) - University of Santa Cruz do Sul, Santa Cruz do Sul, 2012.

OLIVEIRA, L. A.; HENKES, J. A. **Water Pollution:** industrial pollution in the Rio dos Sinos - RS. Florianópolis, 2013. p. 186-221.
OUKI, S. K., KAVANNAGH, M. Performance of natural zeolites for the treatment of mixed metal-contaminated effluents. **Waste Management Reserch,** v.15, p. 383-394, 1997.

REIS, G. S.; LIMA, E. C.; SAMPAIO, C. H. Production of activated carbon from domestic sewage sludge and its application in the adsorption of remazol 5 black dye in aqueous solution. **E-xzacta.** Belo Horizonte, v.8, n.2, p.15-23, 2015.

RIO GRANDE DO SUL (State). DAT-MA Document No. 0616/2008. Geoprocessing Environmental Advisory Unit - River Basins, Rio Grande do Sul, 2008. 13p.

RUTHVEN, **D.M. Principies of adsorption and adsorption process.** United States of America: Wiley - Interscience Publication, 1984. 453p.

SANTOS, C. M. **Use of Orange Peels as an Adsorbent for Contaminants in Water Treatment.** 2012. 126f. (Master's Degree in Environmental Sciences) - São Paulo State University, Experimental Campus of Sorocaba, Sorocaba, 2015.

SHERRY, H. W. The ion exchange proprieties of zeolites. In: lon Exchange, chap 2. New York: Mareei Dekker, 1969.

SILVA, A. P. O. **Comparative study of the adsorption of Cu ions^{2+} and oil in synthetic effluents on hybrid materials based on clay minerals and chitosan.** 2015. 93f. (PhD in Chemistry) - Federal University of Rio Grande do Norte, Natal, 2015.

SOUZA, R. S. et al. Adsorption of chromium (VI) by granular activated carbon from dilute solutions using a batch system under controlled pH. **Acta Amazônica,** Manaus, v. 39, n. 3,p. 661-668, 2009.

YARON, B.; CALVET, R.; PROST, R. **Soil Pollution:** processes and dynamics. Heidelberg, Germany: Springer, 1996. 319p.

Printed by Books on Demand GmbH, Norderstedt / Germany